CONTENTS

PHYSIOLOGICAL INTEGRATION

PHYSIOLOGICAL INTEGRATION

Prepared for the Course Team by Caroline Pond

1.1   Introduction

S324 addresses three major topics: temperature control (Book 2), reproductive physiology (*Essential Reproduction* by Johnson and Everitt) and locomotion (Books 3 and 4). Although these aspects of physiology may appear to be very different, they all involve biological processes that make major demands on animals' energy resources, often competing with each other to do so. Studying them reveals mechanisms that maximize physiological efficiency and adjust the timing and organization of foraging, breeding, migration, etc. to the seasons and other environmental conditions. Neural signals, hormones and other forms of biological 'information' establish priorities between competing organs and tissues and coordinate physiological responses.

Scientists have developed several different ways of 'listening in' as messages pass between tissues. Neural messages can be picked up as tiny, transient changes in voltage using electrodes and electronic recording systems. Signals that take the form of blood-borne molecules such as hormones can be measured chemically, though determining which of the thousands of components of the blood conveys the message, and the origin and significance of changes in its concentration are not always easy. Chemical signals have been intensively studied because, as well as their importance in understanding normal physiology, administering exogenous messenger molecules (or their inhibitors) offers a simple means of intervening to control physiological processes artificially.

However, the messages that eavesdroppers receive may be so incomplete and garbled that they are misinterpreted, leading to misguided and inappropriate intervention. It is very important to study all avenues of communication whether they are easy and convenient to measure or not. Much of *Essential Reproduction* is concerned with the detection and interpretation of blood-borne hormones and other signals that initiate and coordinate gamete formation, conception, pregnancy, birth and lactation.

Most signal molecules work in the nanomolar range of concentration ($nmol\,l^{-1}$ = $10^{-9}\,mol\,l^{-1}$), and many are 'short-lived': they are broken down within seconds or minutes of their formation. Until recently, isolating even very small quantities of most hormones and their binding proteins or receptors was extremely laborious (and for many, technically impossible) but advances in molecular biology and genetic engineering have enabled genes that direct the synthesis of particular molecules to be identified and inserted into bacteria which then synthesize the protein. The required protein can be isolated intact from the bacterial culture and applied to isolated cells or whole animals. These techniques provide scientists and physicians with synthetic hormones and other messenger molecules more cheaply and in greater variety than ever before, enormously extending opportunities for experimental and therapeutic intervention in signalling pathways.

Detailed orders are not by themselves sufficient to enable soldiers to carry out missions, or cells to perform physiological functions: they also need sufficient supplies and equipment. Another way of understanding the limitations of what animals can do, and the interactions between competing activities, is to study the energy and other resources required. This approach features strongly in Books 2, 3 and 4, and underpins the topics covered in *Essential Reproduction*.

The rate and mechanism of energy utilization in individual biochemical processes are usually constant when studied in isolated cells or tissues in a laboratory. The energy expenditure of whole animals used to be regarded as the sum of metabolism of isolated tissues, adjusted according to whether the animal was resting, running or synthesizing new materials for growth or milk secretion. Storage tissues used to be regarded simply a fuel dump from which competing tissues and metabolic pathways 'helped themselves' as required until supplies ran out. More thorough research using modern laboratory methods combined with studies on free-living wild animals have shown these concepts to be simplifications. Rates of energy utilization in almost all tissues can be extensively adapted to different physiological states (e.g. pregnancy, growth) and to environmental conditions such as cold and starvation. At least in mammals, storage tissues 'manage' energy reserves, establishing priorities between competing tissues and ensuring that appropriate materials reach the cells that need them.

This book aims to outline some of the methods and concepts involved in the study of hormones and other messenger molecules, and the energy used in physiological processes, and to explain how these contrasting approaches can be welded together to produce an integrated picture of how animals adapt to diverse and sometimes hostile conditions.

Physiologists study animals to elucidate basic cellular mechanisms and to describe and explain the adaptations that enable the animals to live in their natural habitat. Knowledge about physiological mechanisms can help us to repair damage or malfunctions in living systems as well as to control or manipulate normal physiological processes. The study is at many levels: molecular, cellular, whole organ and whole animal. So many different physiological processes are happening simultaneously that isolating the molecules, cells or organs and studying them *in vitro* (literally 'in glass') is often the only way to establish that a certain structure is involved in a particular process and to work out the mechanisms involved. The theories so developed are confirmed by measurements made *in vivo* (literally 'in life').

The technical limitations of examining natural metabolism *in situ* (literally 'in position') often make it impossible to elucidate the details of a biochemical mechanism except *in vitro*. Sometimes it is possible to verify that processes observed *in vitro* do indeed occur *in vivo*, by making a few key measurements. The physiologist's dilemma is to be able to identify properties and processes in a particular organ, tissue, cell or molecule while at the same time keeping the whole system intact and in its natural environment.

Physiologists often find it more convenient (and cheaper) to obtain fresh tissue for *in vitro* studies and to experiment with small animals, usually rodents such as rats, mice and guinea-pigs, or rabbits. If the real goal of such research is to advance understanding of large species such as humans or farm animals, then there is the additional problem of establishing whether the processes measured in the laboratory animals are a special feature of those species alone, or whether they are common to all, or at least to related wild species of contrasting size and habits. Therefore, nearly all physiology is to some extent comparative as well as strictly experimental. Comparison between species that have a fair amount in common, either through shared ancestry or by living in similar habitats, but differing in reproductive habits, offers a natural 'experiment' and guides extrapolation from small, easy-to-study laboratory animals to more inaccessible species including humans.

Many different approaches have their advantages for certain kinds of problems, and to assess the validity of the data obtained by them, students studying physiology, as much as research scientists planning experiments, have to understand the strengths and limitations of the techniques. Because the provision of energy is an essential prerequisite for almost all biological processes, this book concentrates on energy utilization and storage, but several other topics would illustrate the same principles.

1.2   Measuring energy in biological systems

Energy in its many forms, such as heat, light, electricity, chemical energy, movement etc., plays a central role in the formation, maintenance, activities and disintegration of all living organisms. Most of the energy used by animals is chemical energy released by digestion and metabolism of the complex molecules of food, but small amounts are absorbed directly as heat. All biologists are familiar with the basic similarity between the release of energy from complex organic substances by respiration and by combustion: in both processes, organic molecules are oxidized to form water, carbon dioxide and sulphates, nitrates and phosphates. In combustion, nearly all the chemical energy is released as heat. In respiration, much of the energy is harnessed to the formation of ATP in mitochondria, though the process is never perfectly efficient, so some (at least 25%, sometimes much more) is released as heat. Energy in the form of ATP mediates a huge range of processes, including the synthesis of additional and replacement tissues, muscle contraction and the active transport of substances into and out of cells.

It is, of course, theoretically possible to estimate the maximum possible rate of energy utilization by isolating key enzymes and measuring their maximum activities. However, most enzymes are modulated *in vivo* by hormones and the concentrations of substrates and products, and probably rarely operate at maximum activity. The actual energy expenditure *in vivo* of particular processes, such as the energy used in the synthesis of certain substances or the performance of certain actions, is usually more difficult to measure. Nonetheless, such information is essential for assessing capacities such as how far an animal can travel without stopping to feed, or how long it can fast. Understanding an animal's physiological capabilities may help to explain why certain kinds of behaviour, or certain biochemical pathways, occur more often than others.

Standard chemical methods can be used to measure how much energy would be available if all the molecules in a food item were completely digested and absorbed. However, most diets, particularly those of wild animals, are chemically very complex, and many foods are incompletely digested, so that a meal's metabolizable energy content, the net amount extracted by the animal's digestion, is often substantially less than its chemical energy content. It is usually impossible, except in the case of very simple foods such as pure sugars, to calculate exactly how much energy an animal actually obtains from its food. Accurate measurements of the energetic cost of the biosynthesis of the complex lipids, proteins and carbohydrates that build the animal's body, or the total mechanical work performed in movement are also difficult. One reason is that there are large differences in the efficiency of conversion of chemical energy into mechanical work. The process has been studied most thoroughly in skeletal muscles, where it is found that factors such as temperature, initial length of the muscle, and a host of others alter the relationship between chemical energy consumed and mechanical work performed (see Book 3, Chapters 5 and 6). The efficiency of conversion is at best only about 25%, and can be below 5%, with most of the rest of the energy being released as heat.

◼ From your own experience of lifting and carrying heavy, awkward loads, what happens to this heat?

It warms the body, sometimes very rapidly. Continuous strenuous activity may be welcome as a means of 'getting warm' in cold weather, but unless the excess heat can be quickly dissipated, the body may become so uncomfortably hot that we have to stop.

How to maintain more or less constant body temperature by adjusting heat loss to heat production in discussed in Book 2. The facts that a great many aspects of metabolism involve heat production, and body temperature is precisely regulated, are the bases for the use of heat output as an indirect measure of the rate of energy utilization. This method is widely used for isolated organs, small animals, and micro-organisms, but for larger animals measurements of heat production, although technically quite simple, cannot provide more than an approximate value for total energy expenditure. As well as differences in energy conversion efficiency, the additional heat produced locally is quickly dispersed throughout the body by the blood, and so becomes difficult to measure.

Another limitation is the animal's capacity to tolerate more than small changes in the rate of heat production. If forced into becoming too hot or too cold, animals actively try to shed or conserve heat, strenuous exercise becomes impossible, and they may die if such conditions persist.

◼ Over what range of temperatures would measurements of heat production be most accurate?

Over a narrow range close to the animal's preferred temperature. Measurements made outside this range are likely to be wildly inaccurate. Book 2 discusses what constitutes a species' preferred temperature, and how it is maintained. Such regulation is essential, especially during energetically demanding activities such as breeding, lactating and strenuous exercise.

Table 1.1   Units used to express energy flux in biological systems and metabolic rate.

1 calorie	= 4.184 J	
$1 \text{ cm}^3 \text{ O}_2 \text{ g}^{-1} \text{ h}^{-1}$	= 5.6 W kg^{-1}	
$1 \text{ cm}^3 \text{ CO}_2 \text{ g}^{-1} \text{ h}^{-1}$	= 7.14 W kg^{-1}	
1 kcal h^{-1}	= 4 184 J h^{-1}	= 1.16 W

For large animals in which metabolism is mainly aerobic, it is usually more accurate to estimate energy flux by measuring total oxygen uptake and/or carbon dioxide formation. Plants, micro-organisms and small animals can be enclosed in a small sealed container through which air is pumped. The difference in the proportions of oxygen and carbon dioxide in the in-flowing and out-flowing gases is measured. Sealed chambers the size of small rooms have also been built to make such measurements on large animals, including cattle and people, but often it is more convenient, and less stressful to the subject, to cover the nose and mouth with a mask that supplies air at a measured rate and collects the exhaled gases. Such methods are so widely used that energy utilization in biological systems is sometimes expressed indirectly in the units in which it was measured (e.g. cm^3 of oxygen taken up) and sometimes, perhaps as a hangover from emphasis on the parallels between combustion and respiration, as calories or other units of heat. However, the use of the joule (J) as the universal unit of energy (work) and the watt (1 W = 1 J s^{-1}) as the unit of power makes comparisons between different forms very much easier.* Table 1.1 shows the conversion factors between these units of energy and the various physiological parameters which are used as indirect measures of energy flow.

1.3   Metabolic rate

The total amount of energy needed to sustain basic life functions, such as digestion, respiration, excretion and biosynthesis of replacement tissues, if the animal is neither too hot nor too cold and is doing nothing in particular by way of active locomotion, is called **basal metabolic rate** (**BMR**) or sometimes 'resting metabolic rate' or 'resting energy expenditure'. Although the concept has been much criticised (mainly because of the difficulty of standardizing the 'resting' conditions), it has proved to be too useful, particularly for comparisons between species and animals of different sizes, to be discarded.

The contribution of different organs to total BMR is not in simple proportion to their relative mass. Calculations that combined measurements of the heat production of living, isolated tissues and organs with the gross mass of tissues indicated that, of the 125 J min^{-1} (= 2.1 W) produced by a sedentary adult rat, 20% came from the skeletal muscles (about half of the body mass, BM), 14% from the kidneys (about 1% of BM), 14% from the nervous system (mainly the brain), 12% from the liver (4–5% of BM), 11% from the skin and 8% from the guts (about 20% of BM). Only 4% came from a specialized heat-producing (thermogenic) tissue, brown adipose tissue (BAT). However, per unit of mass, BAT's heat production can be very high (in most mammals it accounts for less

* The prefixes, k = 10^3, M = 10^6 apply to joules and watts in the usual way.

than 0.1% of BM), and, in contrast to other sources of heat, neural and endocrine controls actively regulate its metabolism, adjusting its activities from almost zero to enough to warm the rest of the body at a rate of several degrees per hour. Book 2 includes much more about BAT and its control.

While there is no denying that the extra 4% of heat supplied by BAT in the adult rat may make a significant contribution to the body temperature of specimens living in cold conditions, most biological heat production is the result of metabolic processes taking place in tissues other than those specialized for thermogenesis, particularly the muscles. BAT is more abundant and more metabolically active in adult mammals during hibernation and in some neonates, but the tissue is very small and often entirely absent in many other species, including some that are native to cold climates, such as polar bears. Although their normal body temperature is as high or higher than that of mammals, localized thermogenic tissue similar to mammalian BAT has not been unambiguously demonstrated in birds.

1.3.1   Adjusting BMR

In some animals, including humans, BMR can be altered, albeit only to a small extent, in response to changes in food availability. Excess energy can be dissipated if there is more energy in the diet than required, and total energy expenditure can be reduced, without proportionately reducing body temperature or physical activity, if there is insufficient food available. The exact mechanism by which these adjustments are achieved is still not well understood, but alteration of body heat production is believed to play a major part.

In experiments to investigate this phenomenon conducted in Holland, it was found that, when laboratory rats were given an excess of highly palatable food, they ate more than usual. Their metabolic rate over a 24-hour period was about $1.9\,W\,g^{-1}$ and their energy reserves, mostly in the form of fat in adipose tissue (see Section 1.4.2), averaged 1 100 kJ. When the rats were starved, their total energy expenditure fell by 59% from its normal level; a fifth of this reduction was due to savings on the energetic cost of ingestion and digestion of food (chewing, gut movements, enzyme formation, defaecation etc.), another fifth was due to a reduction in physical activities, such as walking around the cage, but the largest saving in energy was achieved by a fall in the BMR to about half of its original value. Although they were kept continuously at an air temperature of 30 °C, the rats' body temperature fell slightly, from about 37.5 °C to about 36 °C, and their body insulation was improved by sitting with their fur fluffed up, but these changes were not sufficient to account for all of the decrease in BMR.

One of the most consistent features of such studies is that core body temperature remains almost constant, a condition called euthermia, even though metabolic rate, and hence heat production, change significantly. The rates of nearly all biochemical processes are influenced by temperature, with most chemical reactions proceeding faster at higher temperatures, so a constant body temperature facilitates the integration and coordination of the thousands of metabolic processes that go on all the time in living systems. Nonetheless, many small mammals and some birds, especially newly hatched nestlings, can undergo quite large and rapid changes in body temperatures.

However, as the measurements just described show clearly, such adjustments do not occur in rats. The genus *Rattus* originated in the hot, damp climate of south-east Asia, where food is available throughout the year. Laboratory rats are descended from a semi-domesticated strain that has relied mainly upon human food stores and refuse for millennia.

■   In what kinds of mammals would you expect to find much more extensive adjustments to BMR and body heat production?

Small species that are native to climates with long, cold winters during which food is scarce. The brown long-eared bat (*Plecotus auritus*) is one of the smallest mammals found in Northern Europe, weighing only 7–11 g when adult (compared with 300–500 g for laboratory rats). Between May and October, it eats moths and other large insects that it catches at night on the wing or while the prey are resting on tree trunks. Seasonal changes in the roosting temperature chosen was investigated in bats caught near Aberdeen (57 °N), close to the northern limit of the natural range of this species (Figure 1.1). The bats were placed in a large chamber in which there was a gradient of air temperature from 8 to 44 °C, and the positions they chose over periods of 6 h were recorded.

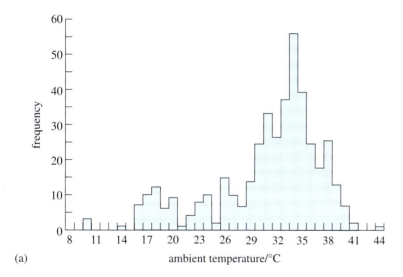

(a)

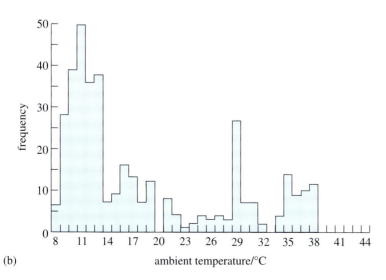

(b)

Figure 1.1   The temperatures selected for daytime roosting by nine brown long-eared bats in (a) summer and (b) October–November. Observations were made every 15 min for 6 h. Data from Speakman and Rowland (1999).

■   What can you deduce from Figure 1.1 about seasonal changes in the bats' choice of ambient temperature?

Bats caught in summer chose to rest in warm conditions, close to 34 °C, but in the autumn, they preferred to be much cooler, at about 10 °C. Roosting bats naturally keep very still, so it is quite easy to measure their oxygen consumption by confining them in a small, closed container similar to that used for rats. Bats placed in such a chamber at 7 °C spent a large fraction of the period in torpor while none of those recorded at 30 °C did so. Torpor is a state in which the body temperature falls to as low as 5 °C and most metabolic processes, particularly those concerned with movement and with the digestion and absorption of nutrients, are greatly slowed, as is the uptake of oxygen needed to fuel these processes. Its mechanism and control are described in Book 2; here we are concerned only with its impact on total energy requirements. Although energy expenditure during torpor is low, large quantities of energy are needed to generate enough heat to bring the body temperature back to normal.

Bats use least energy, and so make the most of the dwindling food supply, by actively foraging at night, then digesting their meal as fast as possible and spending a single long period in torpor each day. Figure 1.2 shows some measurements of the total daily energy expenditure of eight brown long-eared bats kept at 7 °C that spontaneously spent different proportions of the day in torpor.

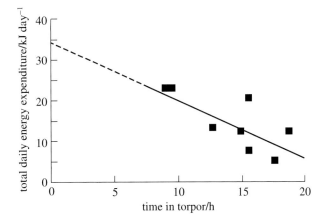

Figure 1.2   Total daily energy expenditure of eight brown long-eared bats kept at 7 °C compared to the time each day that they spent in torpor. The solid line is fitted to the actual measurements, and is extended to form the dotted line that indicates the expected value if the bats were never torpid. Data from Speakman and Rowland (1999).

By selecting a cooler environment and spending an average of 14 out of 24 h in torpor, the wild bats reduce their total daily energy expenditure by one-third over what it would have been had they remained euthermic all the time. This saving enables them to deposit storage lipid in their adipose tissue, even though by October, the number of moths flying around is decreasing so their daily food intake is lower and the effort required to collect it higher.

Many other homeothermic species are able to make similar, and in many cases much greater, adjustments in metabolic rate during periods of food shortages. The mechanisms of temperature regulation, and how major deviations from normal body temperature are implemented and controlled are described in Book 2. Basal metabolic rate and the energy cost of locomotion change in predictable ways with total body mass, for reasons that are described in Book 3.

1.3.2   Non-resting metabolic rate

Metabolic rate increases substantially during locomotion, when an animal is pregnant or growing rapidly and, in the case of homeotherms, when an animal is exposed to extremes of heat or cold. Heat production and oxygen uptake during locomotion can be measured in the same way as BMR. Measuring respiratory gases usually involves training animals to run on a treadmill, the speed of which can be controlled by the experimenter, while wearing a mask for gas analysis. Flying animals are placed in a wind-tunnel, which produces a 'head wind' that can be adjusted so that they remain in more or less the same place while flying at known speeds. Such apparatus has been used to measure oxygen consumed during locomotion at various speeds by animals varying in size from budgerigars, bats and shrews to ostriches and horses (including human athletes).

■   Would the difference between BMR and metabolism during exercise provide an accurate estimate of the rate of energy consumption in the limb muscles?

No, many organs other than the locomotory muscles, such as the heart and the muscles that ventilate the lungs, are much more active during exercise. To meet the additional demands of the exercise, blood flow is also diverted from the gut and liver to the muscles, particularly during strenuous exercise. So, although it is probably correct to say that a fair proportion of the additional oxygen consumed during active metabolism *does* represent the additional energy used by the locomotory muscles, growing tissues etc., it is impossible to determine how much of the extra energy used is supporting increased activity of all the other physiological processes which inevitably take place during growth, locomotion, etc. However, with the proviso that not all the additional energy expended during these 'non-resting' activities is necessarily accounted for *only* by the specialized tissues involved, measurements of active metabolic rate can provide useful estimates of the energetic cost of different activities. For most mammals, the maximum possible metabolic rate is about ten times BMR, but such high values can usually only be sustained for a short period, sometimes only a few seconds. Maximum sustainable metabolic rate is only about 3–5 times BMR.

Measuring oxygen consumption only yields information about aerobic metabolism: anaerobic metabolism, by definition, does not involve the simultaneous uptake of oxygen. Most free-living animals metabolize anaerobically for short periods when oxygen is unavailable, as in diving, at the time of birth, before the fetus starts to breathe air (as you will study in *Essential Reproduction*), and when oxygen cannot reach the tissues fast enough to meet the animal's needs, as in strenuous exercise. For several minutes after the strenuous exercise or a dive has ended, additional oxygen, called **recovery oxygen**, is taken up.* Much of this extra oxygen is used to complete the oxidation of substrates that were only partially oxidized by anaerobic metabolism, but some is consumed by regulatory biochemical processes that maintain the energy-supplying system in a state in which it can respond to rapid increases in demand.

---

* This process used to be called 'oxygen debt', now an obsolete term.

To obtain an accurate picture of total energy utilization, it is important that measurement of oxygen consumption is continued for a long enough period to include uptake of recovery oxygen. The contribution of anaerobic metabolism to the total energy supply can confound attempts to measure the maximum metabolic rate during brief periods of strenuous activity, particularly in animals such as fish, reptiles and amphibians, which support brief periods of strenuous exercise almost entirely by increasing anaerobic metabolism; their uptake of oxygen *during* exercise does not necessarily rise, and can sometimes actually decrease.

1.3.3   Metabolism during growth and healing

Forming new cells, repairing existing ones and removing the debris from those that are beyond repair involve the synthesis of new lipids and proteins and thus produce heat. Growth is usually too slow and continuous for the energy it requires to be separated from basal metabolic rate but transient changes in metabolic rate following accidental and surgical injuries were among the first topics to be studied when the appropriate technology became available. The healing process begins within minutes of tissues being injured, and within a day it is sufficiently extensive to be accompanied by detectable increases in oxygen uptake and heat output. Human patients recovering from surgery or accidental wounds often show a slight increase in body temperature.

Techniques of surgery and anaesthesia have improved so much that procedures are now performed on infants, the very elderly and people who are suffering from wasting diseases such as cancer that would not have been attempted 40 or 50 years ago. Although nearly all now survive the surgery itself, healing and full recovery are often delayed in the elderly and infirm. The effect of undernourishment on the body's capacity to mount a normal metabolic response to healing has been studied in rats. Table 1.2 shows the energy intake, storage and expenditure during the four days after an abdominal operation. The underfed rats were allowed only 48% of the average amount of food eaten by the free-feeding rats for three weeks before surgery and for four days afterwards. At the time of surgery, these underfed rats had lost 36% of their original body mass, including over 82% of their adipose tissue mass.

Table 1.2   Energy balance in rats over a period of 4 days following surgery. Data from Emery *et al.* (1999).

	Free-feeding ($n = 9$)		Underfed ($n = 8$)	
	Surgical	Control	Surgical	Control
Energy intake (kJ/4 d)	623	614	278	278
Energy stored (kJ/4 d)	−238**	−26	−76**	−13
Energy expenditure (kJ/4 d)	874**	640	355**	291

Mean values were significantly different from the corresponding control group: **$P < 0.01$.

■ From Table 1.2, did the operation increase energy expenditure in the lean, underfed rats?

Yes, though by a smaller amount than in the free-feeding rats. Energy expenditure over 4 days increased by (874–640)/640 × 100 = 37% in the free-feeding group and by (355–291)/291 × 100 = 22% in the underfed rats. Evidently, this metabolic response to injury is given 'high priority' because it occurs, albeit in an attenuated form, even when the animals' energy resources are severely stretched.

■ In view of these changes in BMR, would injury cause weight loss?

Yes, especially if appetite decreases as well as BMR rising. In the experiments just described, the free-feeding rats lost an average of 4.2 g and the underfed rats only 1.2 g of fat compared to the controls, but the latter represented 30% of the remaining fat reserves, as against 21% in the well-nourished animals. The ability to partition limited resources between competing claims is clearly essential to dealing with the hazards of food shortage combined with the risk of accidents and encounters with predators that most wild animals experience quite frequently.

Many other pathological conditions, from infection with parasites and pathogens to endogenous changes in hormone levels and brain function produce changes in metabolic rate, though in most cases the exact mechanisms involved are not well understood.

■ How would (a) decreased (b) increased metabolic rate sustained over many months affect appetite and sensitivity to heat and cold, assuming no other changes take place?

(a) Lowering metabolic rate reduces appetite (though people who continue to eat as normal may gain weight unusually rapidly) and reduces tolerance to prolonged exposure to cold. (b) Higher metabolic rate increases appetite (and may lead to extreme leanness unless fully satisfied) and susceptibility to overheating.

All forms of growth entail a high metabolic rate. The additional energy expenditure is supported by increased food intake, or by drawing on energy reserves, often both. A growing fetus *in utero* also has a high metabolic rate, supported by nutrients obtained from the mother's blood via mechanisms to be discussed in *Essential Reproduction*. The fetal body temperature is normally slightly higher than that of adults, so the additional heat so generated flows to the mother. In hot climates, pregnant women (and cows, sows etc.) may become dangerously overheated if they are unable to dissipate the excess heat fast enough. Like other metabolic processes, milk synthesis and secretion also generate additional heat. Modern, highly bred pigs produce such large litters, and the piglets consume so much milk to sustain their remarkably rapid growth that lactating sows easily become overheated. The 'comfortable' air temperature for the young piglets is dangerously hot for the sow, so special pens have to be constructed that maintain mother and offspring at different temperatures.

Summary of Sections 1.1, 1.2 and 1.3

Basal metabolic rate (BMR) is the overall energy consumption of a sedentary, non-torpid animal in a thermally neutral environment; the active metabolic rate includes the additional expenditure of energy on locomotion or heat production in a cold environment. Whole-body energy expenditure can be measured as oxygen uptake or carbon dioxide production, or as heat produced or work performed, but gas exchange is usually more accurate except when anaerobic metabolism makes a significant contribution. Some organs use much more energy, in relation to their size, than others. Food shortage and many normal and pathological states alter BMR for hours or days, leading to changes in body temperature and energy requirements.

Total energy expenditure increases greatly during exercise, mainly due to muscular activity. Growth and healing also entail higher basal metabolic rates, which may lead to higher body temperature, and the response cannot be fully extinguished even by severe food restriction.

1.4   Energy storage and utilization

The highest demand for energy often occurs when animals are eating less, or in some cases not feeding at all. For example, during the breeding season, male animals such as deer are too busy fighting each other and mating with transiently receptive females to devote much time to feeding. Sick or injured animals may be unable to find food and/or appetite is actively suppressed. In these and many other circumstances, animals utilize their energy stores.

Most metabolic fuels, including glucose and long-chain fatty acids, are harmful to many tissues at high concentrations, interfering with membrane function and causing structural alterations in proteins. Glucose is a small molecule (M_r 180) and highly soluble, so it diffuses quickly among proteins. At high concentrations, glucose combines with proteins by a process called non-enzymatic glycosylation, which greatly alters their structure and properties and renders some enzymes inactive.

Animals metabolize continuously but (except in the case of internal parasites) feed only intermittently and so, since it is not physiologically feasible to maintain high concentrations of fuels in the circulation, they convert some of the ingested nutrients into harmless substances that are stored in their body tissues for use 'between meals' or during longer periods of fasting. In many organisms, including most mammals and birds that have been studied, an insoluble polymer of glucose called **glycogen** is the principal short-term energy store. Such biochemical transformation inevitably involves some energy utilization, but only about 5% of the metabolizable energy of glucose is dissipated when glucose is polymerized to form glycogen and reformed by breakdown of the storage molecule. In vertebrates, glycogen is stored mainly in the liver and in the skeletal muscles, but in adults the quantities are always quite small compared to those of lipids, because the molecule binds a lot of water, making it both bulky and of high density.

In many insects, some fish and most terrestrial vertebrates, the chief energy stores are lipids, mostly triacylglycerols (TAGs) that consist of three long-chain fatty acids linked by ester bonds to one molecule of glycerol. TAGs can be synthesized directly from dietary lipids or from glucose, but about 20% of the metabolizable energy of glucose is dissipated by the metabolic pathways that convert it into fatty acids. With the exception of certain parasites, and perhaps some mammalian fetuses, animals cannot synthesize glucose from fatty acids (though plants do so readily, as when building seedlings from seed oils). However, the glycerol component of the TAG molecule can be converted back into glucose and hence can supply energy to those cells (such as certain brain, kidney and red blood cells) which cannot metabolize fatty acids for energy production.

Almost all the fatty acids in storage TAGs contain at least 12 carbon atoms, and most have 16 or 18, so TAGs are relatively large molecules, with a molecular weight about four times that of glucose. Their large size prevents them from permeating membranes and their high affinity for each other enables them to accumulate into highly concentrated droplets that do not interact with other cellular components. Triacylglycerols can thus be safely stored in large quantities until required. Before entering the energy-releasing metabolic pathways, they must be broken down into fatty acids and glycerol in a reaction called lipolysis. Controlling the concentrations of the different fuels in the blood, their utilization by active tissues and storage of any excess involve several hormones and various sensory and neural mechanisms that are discussed in this and later books.

Oxygen is essential for aerobic metabolism but it is also harmful to tissues at high concentrations: it forms highly reactive ions, sometimes called free radicals, that oxidize proteins and lipids, thereby altering their properties, so it is as important to prevent excesses of oxygen as it is to avoid shortages of it. As you will find from the study of air-breathing vertebrates that dive (e.g. ducks, seals and whales, see Book 2), storing oxygen is even more difficult than storing metabolic fuels. Even the most highly adapted divers such as seals and whales cannot sequester enough oxygen to sustain more than an hour of moderately strenuous activity.

1.4.1   Alternative fuels

Diets include a variety of substances and, at least in more complex animals such as insects and vertebrates, the major nutrients are to some degree interconvertible: simple carbohydrates can be synthesized from proteins, and certain lipids from carbohydrates. It is often useful to know what kinds of fuel are being utilized. The ratio of carbon dioxide released per unit of oxygen consumed, called the **respiratory exchange ratio (RER)**, formerly known as the respiratory quotient (RQ), varies slightly depending upon whether the energy is derived from metabolism of carbohydrates, lipids or proteins. Thus, it is possible, in principle, to obtain information about what kind of, as well as how much, fuel is being used by collecting samples of the inhaled and exhaled gases and analysing their composition. Although this method is simple, rapid and can be applied to a conscious, freely moving subject, its interpretation is difficult.

Aerobic breakdown of carbohydrate produces an RER of 1.0, but metabolism of pure lipids, which are rich in carbon and hydrogen atoms but low in oxygen, uses more molecular oxygen, giving an RER of 0.71. The RER for proteins, and for the breakdown of a mixture of fuels, is between these values. In resting people on a normal mixed diet, the RER is usually 0.85–0.95. In spite of the difficulties of its interpretation, RER offers a rapid, simple way of estimating the mix of fuels being consumed with minimal interference with the subject's normal activities. The RER does not alter calculations of total energy released by more than 6%, so differences in the kinds of fuel utilized may be disregarded for calculations of total energy expenditure. However, accurate measurements of RER over a period of hours, together with simultaneous measurements of metabolic rate and key metabolites, can provide insight into the roles of food intake and energy expenditure in determining fuel utilization.

Figure 1.3 shows the RER values and the concentrations of glycerol, insulin* and glucose in the plasma of venous blood of young adult humans during a 24-hour period. On one occasion, the subjects exercised for 1.5 hours, then rested in bed and ate meals consisting of bread, jam and skimmed milk, as indicated. On another visit to the laboratory, the subjects followed the same protocol but did not exercise.

■ Why does (a) RER (Figure 1.3a) decrease immediately after exercise and (b) blood insulin (Figure 1.3c) and blood glucose (Figure 1.3d) increase after meals?

(a) The fall in RER means that more lipids, probably mostly free fatty acids, are being oxidized. (b) The test meals were very rich in carbohydrate. The sugar in the jam would be quickly hydrolysed to glucose in the stomach and would be absorbed, causing a rapid rise in blood glucose, which stimulates insulin release.

In general, these data accord with well-established physiological processes. Consistent with conclusion (a) is the higher blood glycerol recorded in Figure 1.3b, which arises because more triacylglycerols are being hydrolysed into fatty acids and glycerol.

More surprising were the details, highlighted by the effects of the meals and the differences between exercising and sedentary subjects. So many metabolic processes contribute to these measures that it is difficult to attribute any changes with specific causes. However, these data are detailed enough and accurate enough to help us to choose between alternative interpretations. RER remained lower, and blood glycerol slightly higher after exercise than in the controls for at least 24 hours after the activity ended. Blood glucose increased more after meals in the exercised subjects than in the controls, although the concentrations of insulin were similar in the two groups at all times.

■ Could the low RER (Figure 1.3a) be due to the high blood glucose (Figure 1.3d)?

No. High blood glucose should raise the RER, because more carbohydrate would be oxidized. It is not easy to account for the relationship between glycerol and glucose concentrations either. Both lower RER and higher blood glycerol indicate that the period of exercise has altered the mix of fuels used to sustain

* Insulin used to be measured by the hormone's biological activity, as assessed by its physiological effect when tested on laboratory animals, and expressed as 'Units (U) ml^{-1}', rather than its chemical concentration. Improved chemical methods of measurement are now accurate enough for it to be expressed as mol l^{-1}.

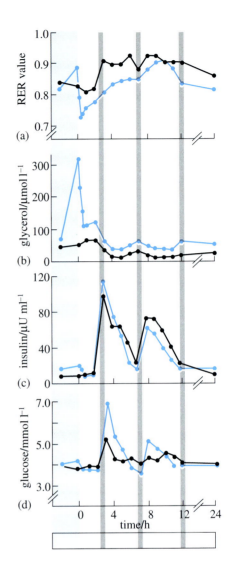

Figure 1.3   Changes in (a) the respiratory exchange ratio (RER), and the concentration in the blood plasma of (b) glycerol, (c) insulin and (d) glucose, in eight healthy people. Meal times are indicated. Results from people when they exercised are shown in blue, and results from when they did not exercise are shown in black. The blue shading indicates the period of exercise, the grey vertical bars, meal times and the points, sampling times. (N.B. No measurements were made until 12 h after the third meal, so its effects on glucose and insulin were not recorded.) Data from Maehlum *et al.* (1986).

metabolism during rest. Because the tissues are oxidizing more fatty acids, they use less ingested glucose, so more of it remains in the blood for longer after meals (Figure 1.3d), regardless of the insulin levels. The greater abundance of glucose in the blood may facilitate the replenishment of glycogen in the muscles, which had probably been depleted during exercise.

■   Do the data reported in Figure 1.3 suggest anything about where the additional glycerol and fatty acids come from?

No. 'Adipose tissue' is clearly a good guess, but these minimally invasive methods can tell us nothing about where the metabolites come from or what controls their release. Further experiments using methods similar to that used for Figure 1.3 showed that this adjustment in metabolic rate and RER continues for many hours after exercise has ended, during which time up to 5.5 W of additional energy is expended, or about 5% of the BMR measured in sedentary people.

■   Does the experiment described in Figure 1.3 suggest that exercise is beneficial for slimming?

Yes. Both during and for many hours after exercise, there is an increased rate of oxidization of lipids, as indicated by the lower RER. Most of the lipids would be derived from lipolysis of adipose tissue TAGs.

1.4.2   Adipose tissue *in vitro* and *in vivo*

In most terrestrial vertebrates, TAGs are found mainly in a specialized tissue, **adipose tissue**, sometimes called white adipose tissue to distinguish it from BAT. Adipose tissue is a tissue unique to vertebrates that is specialized to lipid storage and its regulation. For reasons still not completely explained, adipose tissue in fish, amphibians and reptiles is confined to a few large depots, usually located in the abdomen, where they can undergo large changes in size without impeding adjacent tissues. But that of mammals, and to a lesser extent birds, is partitioned into a few large, and numerous small, depots widely scattered throughout the body. Unfortunately, in rats and mice, only a few depots are big enough to provide enough tissue for analysis, so most of what we know about the basic biology of adipose tissue comes from the study of these depots.

When magnified 100 times, by far the most conspicuous components of all adipose depots are **adipocytes**, which are large, approximately spherical cells embedded in a mesh of extracellular collagen as shown in Figure 1.4. There are also smaller quantities of blood vessels, fine nerves and collagen-secreting connective tissue cells. Adipose tissue is always perfused by a diffuse network of fine blood vessels rather than from a few large arteries and veins, often sharing its supply with adjacent tissues.

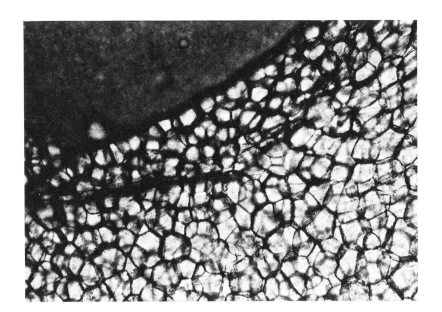

Figure 1.4 Photograph of an unfixed, unstained section of adipocytes of an adult rat. The adipocytes are approximately 80 μm in diameter and contain so much lipid they are almost transparent. The dark object at the top left of the picture is the popliteal lymph node enclosed in the adipose tissue. It contains numerous small cells that cannot be distinguished unless stained and viewed under higher magnification.

Nineteenth-century microscopists used lipophilic (fat-loving) dyes and other stains to demonstrate that adipocytes consisted mainly of a single droplet of lipid, surrounded by a thin rind of cytoplasm and a small nucleus, all encased in the cell membrane. Adipocytes were much larger in well-fed, large animals, in the range of 0.7–3.0 nl (1 nl = 10^{-9} l) in volume, compared to 0.05–0.50 nl in lean, small specimens, mainly because of expansion of the lipid droplet. Investigators also extracted the lipids from macerated adipose tissue using solvents such as ether and chloroform, and found that the lipid content ranged from 70–85% by weight extractable lipid in obese animals to as little as 25% lipid in very lean specimens. Such measurements on dead tissue created the impression that adipose tissue was a passive vessel for lipid storage.

By the mid-20th century, techniques were developed for keeping isolated organs, or fragments of tissues such as liver and muscle, alive for a few hours. As soon as they were excised, the tissues were chopped into small pieces about 1 mm in diameter and placed in a solution containing glucose and salts at appropriate concentrations. The flask was maintained at 37 °C and air or oxygen were bubbled through it. With such apparatus, biologists could measure the rates of uptake and release of metabolites into the solution (e.g. glucose and/or fatty acids) and into the atmosphere above it (e.g. oxygen uptake and carbon dioxide release), although for the latter, much more sensitive apparatus was needed than in the case of measurements of BMR of whole animals.

■   What aspects of the *in vivo* situation were (a) almost perfectly simulated *in vitro*, and (b) barely adequately simulated *in vitro*?

(a) The temperature was probably almost ideal, especially for adipose tissue taken from inside the abdomen, which is almost always at 37 °C (except in hibernating mammals). (b) Chopping up the tissue and shaking the flask help to bring the tissue in contact with the dissolved oxygen and other nutrients, and to remove carbon dioxide, but these procedures often damage cells and cannot fulfil all the functions of normal blood.

It was found that fragments of adipose tissue took up significant quantities of oxygen and glucose, and released carbon dioxide, albeit at a much lower rate than pieces of liver or muscle, and readily incorporated fatty acids into triacylglycerols, proving that adipose tissue has the metabolic capacity to utilize energy, as well as merely to store it. However, adipocytes have very little cytoplasm and few mitochondria, so even when adipose tissue is massive, their metabolism contributes little to the metabolic rate of the whole animal.

This method of studying living processes *in vitro* can be further refined by separating the adipocytes from the extracellular material and other kinds of cells. Separating different types of cells is usually technically tricky, and often causes damage to many cells, but two important properties of adipocytes make them exceptionally easy to isolate. The cells are encased in extracellular collagen, which can be selectively broken down by specific enzymes, and they float in water because they contain a large proportion of lipid. If small pieces of freshly excised adipose tissue are incubated in a solution containing collagenase at 37 °C for about an hour, the adipocytes float to the surface, leaving the connective, neural and vascular cells at the bottom of the flask.

Such *in vitro* methods have been extensively used to study how adipocytes respond to various agonists, and to investigate the mechanisms of triacylglycerol formation, lipolysis and fatty acid transport into and out of cells and in the blood. Water-based enzymes do not readily bind to lipids, especially when they are concentrated into droplets, so lipid-binding proteins are necessary to facilitate lipolysis of the TAGs in the lipid droplet. Once released from the storage cells, the long-chain fatty acids, which are almost insoluble in aqueous blood plasma, circulate in the blood attached to extracellular binding proteins, many of which are much larger molecules than the fatty acids they escort.

1.4.3   Adipocytes respond to signals

Insulin is one of the most thoroughly studied hormones involved in the regulation of metabolism of lipids and glucose in the liver, skeletal muscle and many other tissues. Its main effects on white adipocytes isolated from a large depot of the rat are illustrated in Figure 1.5. The rate of uptake of glucose and the rate of lipolysis, as measured by the release of glycerol are reciprocally altered by applying insulin.

■   Are the adipocytes equally sensitive to insulin over the range of concentrations tested?

No. The rate of change in response is much greater at concentrations between 1 and 100 $\mu U\ ml^{-1}$. The normal concentration of insulin in the blood plasma of fasting adults is about 10–15 $\mu U\ ml^{-1}$ (50–500 $pmol\ l^{-1}$). In humans and other mammals that have been studied, values outside the range of 5–130 $\mu U\ ml^{-1}$ are always accompanied by serious malfunction in many tissues. Very high levels of insulin greatly reduce blood glucose concentrations and so lead to heavy sweating, fits, unconsciousness and many other less obvious symptoms, while too little insulin produces the classic symptoms of diabetes, and rapid breakdown of TAGs and of protein from the muscles and the liver causing glucose and other relatively large molecules to be excreted through the kidney. If such conditions persist for more than a few minutes, the nervous system or the kidneys may be damaged irreversibly. It is not surprising, therefore, that insulin

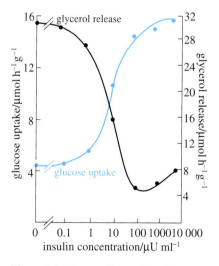

Figure 1.5   Some effects of insulin on glucose uptake and glycerol release on isolated rat adipocytes. The blue line represents glucose uptake and the black line, glycerol release. Data from Green and Newsholme (1979).

has most effect on the metabolism of the adipocytes within the normal physiological range of concentrations. Any responses to insulin outside this range must be regarded as non-functional properties of adipose tissue that tell us little or nothing about its normal physiological role.

The response of isolated adipocytes wanes after a couple of hours in high concentrations of insulin but the change is completely reversed by incubating them in normal medium. Brief exposure to concentrations of hormones or metabolites far outside the physiological range does not cause any detectable permanent damage, but the functioning of small blood vessels and of delicate tissues such as the eye and the kidney that use energy at a high rate are often impaired after many years under abnormal endocrine conditions, as happens, for example, in very obese people, those who smoke or drink heavily, and those suffering from diabetes or other chronic metabolic disorders.

The mechanisms involved in cells' responses to chemical messages are easily studied once the appropriate molecules have been identified and synthesized or extracted from natural sources in usable quantities. Finding out where and how the messenger molecules arise calls for different kinds of investigation. Insulin was among the first hormones for which the source was established. It had been known since the late nineteenth century that mammals in which the pancreas is surgically removed immediately became diabetic. Like humans who develop diabetes, such animals quickly became severely emaciated but could be kept alive for several weeks on a suitable diet. In 1922, two young Canadian physicians, Frederick Banting and Charles Best succeeded in correcting the major symptoms of diabetes in a dog whose pancreas had been removed by repeated doses of material extracted from the pancreas of a second dog, thus demonstrating that the basic cause of disease is the lack of a soluble secretion from this gland.

This simple experiment produced such clearcut results because, as subsequent research showed, insulin is synthesized only in the islet cells of the pancreas. We still don't know exactly why other kinds of mammalian cells cannot produce this small peptide, but attempts to coax them into doing so (by artificially inserting the appropriate genes) have so far met with only limited success.[*] In this respect, insulin is unusual. Since these pioneering investigations, scores of circulating messenger molecules have been identified, the majority of which turn out to be synthesized and secreted by more than one kind of cell, or by several regions of a single organ such as the brain or gut (often both). Determining their sources requires more sophisticated techniques.

Several other hormones, including adrenalin and glucagon, the neurotransmitter noradrenalin and metabolites such as lactate ions, influence the rates of lipolysis and glucose uptake in isolated adipocytes. For example, adrenalin and noradrenalin act on adipocytes during exercise and other forms of physiological stress including anxiety. Their effects on lipolysis and glucose uptake are almost exactly the inverse of those of insulin. Very recently, many **cytokines** have also been found to modulate lipolysis or the uptake of lipids and glucose in certain adipocytes, especially those from minor depots such as the popliteal (Figures 1.4

---

*Insulin for therapeutic use was first produced by extraction from the pancreases of pigs and cattle slaughtered for meat. Since the 1980s, 'human' insulin produced by bacteria engineered to contain the relevant human gene has also became widely available.

and 1.6), as much or more than hormones and noradrenalin. Cytokines are protein messenger molecules that differ from hormones in that they usually act locally, being produced by cells or tissues adjacent to those that respond to them so they may appear in the blood only at low concentrations or not at all. Cytokine activity can often fluctuate over a short time scale, with cells induced to increase or curtail secretion within a few minutes. The messenger molecules (or their receptors) may remain active for any period of time from a few seconds to several days.

■   What aspect of laboratory practice could explain why large responses to cytokines remained almost unnoticed until recently?

Only adipocytes from the large depots were used for laboratory investigation and cytokines are produced and act predominantly in minor depots, which are too small to be studied in rats and mice. Site-specific differences and temporal changes in responses to cytokines were identified using guinea-pigs which are much larger (and fatter) than normal rats, but few genetic mutants are known in this species, and it is not as readily manipulated by dietary change. It turns out that the only depots large enough to be studied in rats and mice are atypical of the others in having minimal response to cytokines. Concentrating all research on these species delayed discoveries about these regulators of adipocyte metabolism, which have proved to be particularly important in energy management during disease and injury (see Section 1.3.3).

Many messenger molecules and metabolites alter each other's action in several different ways: by direct inhibition or stimulation of a biochemical pathway (as in Figure 1.5), by enhancing or reducing the sensitivity of a pathway to another hormone or cytokine, or by affecting the rate of release of another hormone at its sources elsewhere in the body.

■   Which of these processes could be studied *in vitro*?

Inhibition or stimulation of a biochemical pathway and alteration of the sensitivity of a pathway to another hormone. It is always tempting to attribute greater importance to those processes that can be quantified from studies *in vitro*. In assessing the relevance of such studies to what goes on in the normal intact animal, it is important to remember that conclusions are inevitably led by the availability of techniques to study them. Tissues *in vivo* are subjected to a continuously changing cocktail of hormones, cytokines and metabolites, each of which can act on several different biochemical pathways. Their effects may be suppressed or even completely abolished by other agents which are not being monitored, and may indeed not even have been identified as affecting the biochemical pathway under investigation.

1.4.4   Receptors determine the response to signals

Hormones and cytokines do not influence metabolism until they are bound to specific receptors located on the cell surface. The abundance of these receptors determines the capacity of the cells to respond to the hormone. Receptors are synthesized and deployed by the receiving cells themselves, and are larger (often much larger) and more complex molecules than the hormones. Isolating and quantifying receptors used to require lengthy chemical procedures that only worked with quite large quantities of starting material. Such methods showed that many tissues including muscle and adipocytes that are exposed continuously to high concentrations of insulin lose their receptors, so they become less responsive to the hormone, as happens in many middle-aged people in the early stages of diabetes.

Within the last decade, methods for producing specific antibodies that bind selectively to particular proteins have been combined with those for making coloured or fluorescent dyes, to produce simpler and more accurate ways of revealing receptors and many other large proteins that are present in low concentrations. Figure 1.6 shows the popliteal adipose depot of a rat's hind leg, enclosing the popliteal lymph node, stained with immunofluorescent antibodies to receptors to a cytokine called **tumour necrosis factor** α (TNFα). Its name reflects its first discovery as a substance that suppresses tumour growth but it has since been found to have many other roles, mostly involving the immune system. Many kinds of immune cells secrete TNFα and/or its receptors including those that are concentrated inside lymph nodes.

Only a few receptors can be detected on these adipocytes until the lymph node is activated by injecting a key substance isolated from potentially harmful bacteria. But within half an hour of such stimulation, and when stained with immunofluorescent antibodies raised against the receptor for TNFα and viewed under ultra-violet light, the receptors can be seen on the surface of the large adipocytes, as well as in the lymph node itself, as shown in Figure 1.6. At first, they are much more abundant, producing brighter staining, in the adipocytes that enclose the lymph node in a shell approximately 1 mm (= 10–15 adipocytes) thick than in those adipocytes further away.

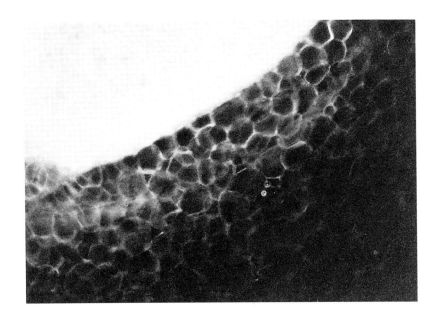

Figure 1.6   TNFα receptors on popliteal adipocytes 30 minutes after the lymph node enclosed therein has been activated by injecting a key substance isolated from potentially harmful bacteria. This photograph shows the same tissues in the same orientation as Figure 1.4 but receptors are made visible when stained with immunofluorescent antibodies raised against them and viewed under ultra-violet light. The adipocytes are approximately 80 μm in diameter, and the lymph node is approximately 2–3 mm in diameter. Courtesy Hilary MacQueen.

■   Comparing Figure 1.6 with Figure 1.4, can you discern any difference between receptor-bearing and other adipocytes when viewed under ordinary light?

No. There are no obvious differences between adipocytes adjacent to the node and those further away from it, although only the former take up the immunofluorescent antibody stain for TNFα receptors.

This study reveals site-specific differences not previously suspected from studying the tissue with normal histological methods. Techniques such as that used for Figure 1.6 have shown that adipocytes can synthesize and deploy a wider range of receptors than was previously believed, and are equipped to respond to signal molecules that until a few years ago were thought to be restricted to the immune system and other highly specialized types of cells.

Certain adipocytes are equipped to respond to various locally acting cytokines as well as to blood-borne hormones and neurotransmitters, indicating that local control of their activities may be much more important and widespread than measurements using adipocytes *in vitro* had suggested. It was also found that most receptors are much more transient than anticipated, appearing and disappearing from cell surfaces and undergoing large changes in abundance within hours or minutes according to physiological conditions. Receptors for most cytokines can be found on a wide variety of different kinds of cells, including, as in Figure 1.6, the numerous lymphocytes and macrophages inside the lymph node. To make things more complicated, most cytokines, including TNFα, turn out to have more than one kind of receptor which may occur in different combinations on different cells, and the abundance of each changes on its own time course.

■   Why would techniques such as those used for Figure 1.5 not reveal the site-specific properties of adipocytes seen in Figure 1.6?

Because the isolated adipocytes were all mixed up together, obliterating any site-specific differences that were in the original sample. The biochemical methods measure a total flux of metabolites, and the calculations assume that all cells contribute equally throughout the experimental period (1 hour in Figure 1.5).

Using the lymph node as a marker ensures that exactly homologous samples are located in each specimen. Cutting sections maintains the natural anatomical relations of the cells, thus allowing site-specific differences to be demonstrated by staining with immunofluorescent antibodies. Some degree of site-specific difference in responsiveness has been demonstrated for most known agonists that act on adipose tissue. Hormones may be readily detectable in the blood, but a lack of sufficient receptors may prevent them from acting on some or all of the target cells. A large fraction of the whole-body response may be due to the actions of a minority of cells equipped to respond strongly to the agonists.

1.4.5   Adipocytes send out signals

As well as receiving chemical messages from other tissues, adipose tissue is now known to synthesize some of its own and release them into the blood, thus meeting the most important criterion in the definition of a gland.

The discovery of messenger molecules produced in adipose tissue is one of the most successful outcomes of studies of spontaneous mutant laboratory mice. In the early 1950s, staff at the world's largest mouse-breeding establishment at Bar Harbor in Maine, USA noticed some exceptionally large mice with rapacious appetites, born to apparently normal parents. This trait was inherited as a single gene that was named *ob*, and homozygous mutant *ob/ob* mice became the first 'genetically obese' laboratory animals to be described.

Mutant mice are of average size at birth but they eat more and grow to be three times as heavy as normal mice, the increase being due mostly to the enormously expanded adipose tissue. These observations suggest that the brain mechanisms that relate appetite to fatness were 'unaware' that the adipose tissue was enlarged, leading the mice to eat themselves to obesity. This concept stimulated a search for a blood-borne factor that in normal mice enabled adipose tissue to communicate with the brain. In the early 1990s, a small soluble protein was isolated and was first known as *ob*-protein, before being given the more elegant name of **leptin**, derived from the Greek adjective λεπτοσ (leptos), which means peeled (as of fruit), slender or delicate.

Modern methods for quantifying proteins at low concentrations and for producing them artificially by inserting the gene into bacteria have enormously facilitated the study of such newly discovered messenger molecules. Within a few years, the concentration of leptin could be measured accurately in small blood samples, and leptin synthesized in bacteria could be administered in controlled doses. Such accuracy is essential to obtaining reproducible experimental results. Like most messenger molecules, leptin's concentration in the blood is always very low, of the order of 1–4 ng (10^{-9} g) per millilitre of blood. Following isotopically 'labelled' leptin molecules suggests that they remain in the blood circulation for about 3 hours.

Because leptin was first described as arising from the adipose tissue, and had such a dominant action on appetite, all the early research was directed towards understanding its role in satiety and lipid deposition. If mice are starved for 48 hours, the leptin concentration in their blood falls to about a third of the normal value. Mice eat several times every night, so fasts of this duration make them quite hungry and when refed, they eat substantial but not excessive amounts. If leptin is absent or continuously low (as in homozygous *ob/ob* mutants), mice eat to excess and become obese, but they revert to normal appetite, and eventually to normal body mass, when leptin is injected or its endogenous production restored.

Leptin was soon found to pass from the blood into the brain, and bind to specific receptors on many neurons in the hypothalamus, the region of the brain that controls appetite (and many other drives and emotions). This small protein seemed to have all the properties of the long-sought signal that regulates appetite in proportion to the level of fat reserves in adipose tissue. Obesity could be caused by disruption to these chemical signals from the adipose tissue to the brain, leading to chronic overeating. Excess lipids cannot be excreted, so unless they can be oxidized to produce energy, they have to be deposited in the storage tissues.

These discoveries led to much excitement among pharmacologists, who hoped that synthetic leptin could be used as a slimming drug, suppressing appetite at its source in the brain. But the notion quickly encountered difficulties.

Measurements of the blood concentration of leptin in large samples of people revealed that the amounts are variable, but the obese tend to produce more leptin than those of normal weight.

■   Would you expect this finding?

No. It is the opposite result to what would be predicted from the study of *ob/ob* mice. Human and mouse leptin are chemically almost identical, but in contrast to the genetically obese mouse, very few of the many different forms of severe obesity in people can be attributed to mutations in the gene that codes for it, though in some cases, there may be minor defects in an adjacent controller gene.

In fact, it is still not clear exactly what change in the adipocytes prompts them to secrete leptin. Being very small mammals, mice use energy at a very high rate per unit body mass (more on this topic in Book 3), so at least some of their adipocytes shrink detectably after a day or two without food. The same cannot be said of adult humans, whose glycogen reserves in the liver and muscles are sufficient for at least two days of starvation. Nonetheless, leptin concentration in the blood rises by 40% at the end of a single day of feasting, and falls by a similar amount after a brief fast, without any detectable change in mass of those adipocytes that can easily be sampled *in vivo*.

■   Can you suggest an explanation for this finding?

There could be site-specific differences in leptin secretion, as there are for most other properties of adipose tissue (e.g. Figure 1.6). Demonstrating site-specific differences is not easy, because even under ideal conditions, leptin is produced in such small quantities. But there is some evidence from both mice and people that some adipose depots contribute more than others to the total amount of leptin circulating in the blood.

In the 1970s, another mutant mouse (named *db/db*) appeared that had symptoms almost identical to that of the *ob/ob* mutant mouse, but normal amounts of leptin were found in its blood both after feeding and when fasting.

■   What single gene defect could produce the symptoms of leptin deficiency without affecting leptin production at all?

Defects in the hormone's receptors. Joining the blood systems of an *ob/ob* mouse with a *db/db* mouse (a technique known as parabiosis, see Book 3, Chapter 2) produced results consistent with this hypothesis. Without fully functional receptors, the target cells are unable to respond to leptin, so from a physiological point of view, it might as well not be there. Recent research has shown that leptin (like other hormones) also binds to special binding proteins, thereby becoming unavailable to the receptors. Short-term or long-term changes in receptors and binding proteins as well as in leptin secretion all contribute to its physiological effects.

Leptin was among the first messenger systems to be discovered as a direct result of studying genetic mutants. The *ob/ob* mutation appeared spontaneously, as did that in which leptin production itself was normal but the gene for its receptor was deficient. During the 1980s, techniques were developed that make it possible to produce mice in which in particular genes are altered or deleted, as desired. Such **'knock-out' mutations** offer a powerful new tool in investigating complex physiological systems *in vivo* and are currently being used to study the action of leptin and other hormones and cytokines in cell function and energy partitioning, as well as many other physiological systems. For technical and economic reasons, the method is at present only used in mice, though in principle it could be extended to rats and domestic livestock.

1.4.6   Leptin in other tissues

Thorough study of *ob/ob* mice showed that leptin influences more physiological systems than just food intake. In spite of their size, the mutant mice appear to be starving: as well as being insatiably hungry, they are reluctant to take any exercise and their resistance to infection, injury and disease is weak. They also do not respond appropriately to cold. In fact, they die from hypothermia after a few hours at 5 °C, while normal mice produce more heat that maintains their body temperature (using biochemical mechanisms described in Book 2).

These symptoms resemble the response of genetically normal rats to starvation as described in Section 1.3.1. BMR has been reduced by the absence of leptin, so they produce less heat, and their metabolic rate cannot be raised to support vigorous exercise or tissue growth.

Continual measurements over a week or more of the metabolic rate of laboratory mice treated with daily injections of leptin indicate that this agonist also stimulates metabolic rate, though on a longer time scale than it suppresses appetite. As explained in Section 1.3.1, chronic food shortage prompts various economies of energy expenditure in small mammals, including reduction of BMR. In genetically normal laboratory mice, the change was only 5–15% of normal, so it would have been undetectable except with the sensitive apparatus used.

■   What role might this effect of leptin play in the energy budget of wild mammals?

If more leptin raises metabolic rate, less might reduce it. When continuous food shortage lowers leptin secretion for several days at a time, such an effect would help to eke out the remaining reserves. As yet, we do not know how important such effects are in enabling starving animals to survive longer between meals, but this combination of actions certainly makes sense.

By 1997, it was clear that leptin did more than adjust appetite and metabolic rate to energy stores in adipose tissue.

■   From what you already know of the physiological roles of adipose tissue, on what other physiological processes might leptin act?

Since adequate energy stores are often an essential condition for breeding, normal levels of leptin may be necessary for fertility and successful reproduction.

The homozygous *ob/ob* mutant mice are infertile, with females failing to ovulate[*]. Daily injections of leptin for several weeks restore fertility to these mutant mice, enabling them to become pregnant. Dosing genetically normal

---

[*] The strain is bred from mice heterozygous for the *ob* gene, which are fertile and of near normal body mass.

mice with leptin maintains ovulation even while they are fasting. When given to young female mice that produce sufficient endogenous leptin to maintain their normal body weight, the hormone hastens the maturation of the reproductive system and enables them to breed earlier than the controls.

Very sensitive techniques of molecular biology reveal that the genes for leptin and its receptor are active in the placenta of both mice and women, and in various tissues of the fetus. So this hormone, previously thought to be secreted only by adipose tissue is now found to be synthesized in reproductive tissues. In humans (though not necessarily in much leaner wild mammals) adipose tissue is still thought to be the major source of leptin but the question of how the signal from the reproductive tissues is integrated with that from the adipose tissue is currently being investigated.

■   What do these experiments suggest about the role of leptin in sexual maturation?

Leptin may serve as the signal that lipid stores in adipose tissue are sufficient to give the 'go ahead' for puberty and the onset of fertility. Consistent with this hypothesis was the demonstration of specific leptin receptors on certain cells in the testis and in the ovary.

■   Is leptin's role in reproduction relevant to its exploitation as a slimming drug?

Most certainly yes. Unforeseen side-effects are major hazards for newly developed drugs, and those affecting the reproductive system are among the most difficult to detect because most people only breed a few times in their life.

These investigations suggest that leptin acts much more widely than was thought when it was first identified. More recently, receptors for leptin have been found in liver, skeletal muscle and the islet cells of the pancreas that secrete insulin, as well as on many adipocytes. Leptin is now known to modulate the action of both noradrenalin and insulin on adipocytes (see Figure 1.5), suggesting that it functions as general controller of deposition and utilization of energy reserves.

But there are inconsistencies. For example, although tumour necrosis factor α and other cytokines with similar functions stimulate adipocytes *in vitro* to produce more leptin, and weight loss invariably accompanies chronic diseases such as AIDS (Acquired Immune Deficiency Syndrome) and the terminal stages of cancer, efforts to demonstrate that leptin mediates the loss of appetite and rapid reduction of adipose tissue have so far been unsuccessful. Another mystery is the response of leptin levels to strenuous exercise, the most frequent major drain on energy reserves in most wild animals.

■   Can you suggest two reasons for *expecting* to find a change in leptin secretion after a few hours of exercise?

1. Continuously high levels of leptin are found to stimulate basal metabolic rate, so it may affect metabolism during exercise. 2. Leptin levels can change over a few hours after eating large meals, so abrupt changes on the timescale of bouts of exercise cannot be impossible. Nonetheless, no changes could be detected in the blood leptin of lean young male athletes after they had run a distance of 20 miles.

A carefully planned experiment on the interaction between food availability, appetite and metabolic rate exposed the subtleties of the multiple actions of leptin. Figure 1.7 shows the effects of injecting three widely different doses of leptin on the change in total daily energy expenditure (i.e. metabolic rate averaged over each day) in genetically normal mice given unlimited access to food, and those restricted to about two-thirds of the amount eaten by the free-feeding group.

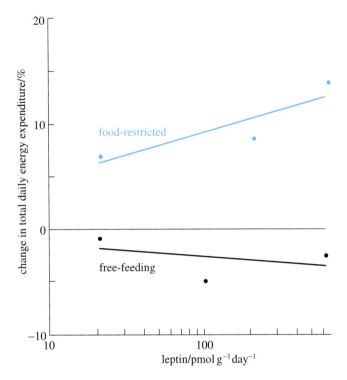

Figure 1.7   The action of leptin injected at three different doses ranging between 20 and 600 pmol (g body mass)$^{-1}$ d^{-1} for 3 days on the change in daily energy expenditure in mice allowed unlimited or restricted access to food. The blue line represents underfed mice and the black line free-feeding mice. Because the doses of leptin used differ by more than an order of magnitude, the x-axis has a logarithmic scale. Data from Döring *et al.* (1998).

■   What can you conclude from Figure 1.7?

The extra leptin induced an increase in BMR only in the food-restricted mice. The experimental treatment hardly changed this measurement in the free-feeding mice.

Continuous monitoring of energy expenditure throughout the day and night revealed that this constancy of BMR was due mainly to a greater decrease in energy expenditure during the animals' rest periods: activities such as foraging, caring for and suckling the young were unaffected. Other measurements showed that, especially at large doses, the leptin caused the free-feeding mice to eat less food, and their body mass declined slightly, but even the very thorough studies could detect no changes in energy expenditure. These data suggest that leptin controls thermoregulatory energy expenditure when food supplies are scarce but acts primarily to change food intake when food is abundant.

Other hormones, discussed in Book 3, also control the partitioning of resources between growth of somatic tissues such as muscle and the skeleton, and reproductive structures, behaviour and secretions. They have been intensively studied because of their applications to livestock farming. Administering hormones (or synthetic molecules that are similar enough to work the same way) can induce larger, more meaty muscles (i.e. more somatic growth) or higher milk yield (i.e. directing resources to 'reproduction') as required.

Summary of Section 1.4

The respiratory exchange ratio (RER) is measured from exhaled and inhaled gases and indicates the chemical composition of the fuel used. Measuring RER in combination with levels of metabolites and hormones in the blood during controlled periods of exercise and meals reveals physiologically significant changes in fuel utilization and BMR after as well as during exercise that can be related to cellular processes. Adipocytes are the largest and most abundant kind of cell in white adipose tissue, and can be isolated by breaking down the extracellular collagen. The effects of hormones and metabolites on these and other processes can be studied *in vitro* using isolated adipocytes. The response depends upon the presence of appropriate receptors that the adipocytes synthesize and deploy themselves. The abundance of function receptors changes with physiological state.

Secretion of messenger molecules from adipocytes was discovered much more recently, initially from the study of spontaneous genetic mutations in mice. Mammalian adipocytes secrete several cytokines and other proteins that could act as messenger molecules. The best characterized hormone is leptin which binds to receptors in the brain to inhibit feeding, and has longer-term actions on metabolic rate. Very detailed measurements reveal that its action changes according to nutritional conditions. In non-pregnant animals, leptin prevents excess weight loss when food intake is low by controlling thermoregulatory energy expenditure and prevents excessive weight gain by limiting appetite when food is abundant. The gonads and other reproductive tissues also produce leptin and/or its receptors, which may coordinate the partitioning of energy reserves between reproduction, growth and other activities. In spite of many major advances in our knowledge of when and where this hormone acts in the body, there are gaps in our understanding of its exact role as arbiter of energy supplies between competing physiological processes.

1.5   The composition of lipid stores

As we have seen, a fall in body temperature is a common means of reducing energy expenditure. But homeothermic animals cannot remain continuously at a lower temperature: growth, locomotion and many other functions are slowed down or in other ways seriously impaired. So mammals and birds need to rewarm as soon as energy supplies allow. Fats solidify when cooled, so is the fuel in a physical state that is accessible to enzymes and can be utilized to generate the heat that restores normal function? The chemical composition as well as the quantity of reserves matter.

When stimulated by the appropriate hormones and neurotransmitters, glycerol and fatty acids are released from adipose tissue (see Figure 1.5) and fuel prolonged exercise such as migration, short periods of fasting between meals and much longer periods during hibernation. The enzymatic processes involved in fasting and starvation are essentially similar to those of slow exercise, but there is a critical difference: during exercise, the body is warm, often slightly warmer than when sedentary, but in hibernation, the body temperature is low, sometimes falling by 35 °C to close to 0 °C. Enzymes do not work on solidified fats, any more than they function in frozen water. Animals must still be able to metabolize the triacylglycerols in their adipose tissue, albeit much more slowly than when they are fully active. The fatty acid composition of triacylglycerols is largely irrelevant to their role as fuel when animals are warm — that is why we are able to utilize any of a wide range of different fats and oils in our diet — but it is crucial for their use during hibernation.

The role of the composition of dietary lipids in hibernation was investigated experimentally in various small mammals in captivity including ground squirrels (*Spermophilus*), rat-sized rodents that hibernate for several months during the North American winter. Figure 1.8 shows that *Spermophilus* remain

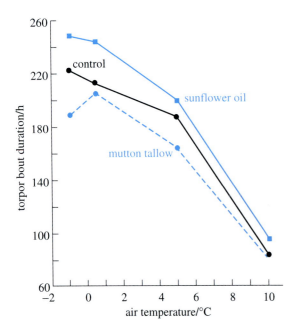

Figure 1.8   The duration of torpor bouts in adult ground squirrels exposed to various air temperatures during natural winter. During the previous three months, the squirrels had been fed on plain rat chow or chow to which had been added 10% of either sunflower oil (which is liquid at room temperature because it is rich in polyunsaturates), or mutton tallow from sheep (which is solid at room temperature and consists mostly of saturated and monounsaturated fatty acids). From Geiser and Kenagy (1993).

in torpor for longer, especially under the coldest conditions tested, and so would be better able to survive long winters, when plenty of polyunsaturated lipids are included in their diet during the weeks preceding hibernation than when they are fed on saturated fats of similar energy content. One reason could be the contribution of unsaturated fatty acids to lowering the melting point of the triacylglycerols and the structural phospholipids in membranes. The lipids remain more fluid, and hence retain their proper affinity for carrier molecules and the capacity to bind to enzymes, at cooler temperatures. The chemical composition of the storage lipids, together with other aspects of adipose tissue, is thus adapted to the physical conditions and its role in whole-body metabolism.

■   Under what environmental conditions recorded in Figure 1.8 does the composition of the diet have the greatest effect on hibernation?

In cold conditions below 0 °C the squirrels remained torpid for $(250-190)/190 \times 100 = 32\%$ longer after eating sunflower oil compared to eating mutton tallow. Metabolic rate during hibernation is strongly influenced by body temperature, so hibernators use far more energy when inappropriate fatty acids in the adipose tissue make the process inefficient.

■   Is the control diet of plain rat chow ideal food for this species in autumn?

No. Adding a little sunflower oil enables the squirrels to sleep for longer, with torpor bouts lasting over 10 days at and below 0 °C.

As will be discussed in Book 2, hibernation is an active, physiologically controlled process. Especially in mammals, metabolic preparations can be identified days or weeks before the animal actually starts to cool. As the weather becomes cooler and the days shorten in autumn, squirrels and other hibernatory rodents actively seek seeds, nuts and other foods that contain these lipids. The woodchuck or marmot (*Marmota flaviventris*) selectively retains linoleic acid (a common polyunsaturated fatty acid with two double bonds among the 18 carbons) before hibernation: the saturated fatty acids are released and oxidized by the muscles, liver, etc. while the animal is warm and active, but the polyunsaturates remain in the adipose tissue, for use when the body is cold. The obvious conclusion is that these herbivores depend upon the adipose tissue's ability for selective retention of triacylglycerols containing certain fatty acids, and upon the increased availability in autumn of seeds rich in polyunsaturated lipids. Failure of a seed crop could prevent efficient hibernation and thus lead to the animal's death from starvation, even if plenty of other foods were available.

As well as acquiring storage lipids of appropriate fatty acid composition, the biochemical mechanisms for mobilization of the reserves, including binding proteins and enzymes, must be adapted to function at low temperature. The affinity of fatty acid binding proteins for various fatty acids from rats were compared *in vitro* with those of ground squirrels. It was found that hibernators' proteins bound the lipids as efficiently at 5 °C as at 37 °C, while those of rats (which do not hibernate) were effective only at the higher temperature.

1.6   Conclusion

Effective partitioning of energy between competing activities is essential for animals to thrive and breed in their natural environment. This chapter has concentrated on energy reserves and expenditure because they are easy to measure, but a similar picture would emerge from the study of other limiting nutrients or resources. Modern laboratory techniques make separating and identifying large biological molecules so much easier than it used to be that the number and variety of messenger molecules and communication systems are growing very fast. It is now much easier to characterize a protein and isolate its gene than to establish what it does in the living animal.

Demonstrating natural function involves detailed studies of living systems at organ, tissue, cellular and molecular level, but in evaluating such data it is essential not to lose sight of the natural context of the processes elucidated. It is nearly always possible to measure metabolite concentrations more accurately from excised tissues *in vitro*, and the use of isolated cells permits identification of particular physiological processes with certain cell types. However, conditions *in vivo*, particularly blood perfusion, cannot always be simulated precisely *in vitro*, and it is rarely possible to determine exactly how excision and isolation alters the metabolism of the cells. The use of natural and directed gene mutations in mice provides a new tool for investigating physiological processes *in vivo*. Physiology is always a synthesis of investigations at several different levels and using many different techniques.

Mammalian adipocytes are now known to secrete several cytokines, including TNFα (Figure 1.6), and certain other proteins that could act as messenger molecules but, as yet, we know little about what the signals might mean, or when they are deployed. But their discovery shows that at least in mammals, adipose tissue is much more than just a repository for storage lipid. It is actively involved in communications between tissues and in resource management, directing nutrients to where they are required in appropriate quantities.

It is clearly impossible to study all biochemical processes in all species; certain species, often small, rapidly breeding forms such as rats and mice, prove to be more convenient for certain techniques, and so are studied intensively. From the mass of data thus accumulated, we build up a text-book picture of 'standard' mammalian physiology and biochemistry. However, although basic anatomical structures and physiological mechanisms are similar in animals that share a common ancestry, detailed studies of wild species in their natural habitat reveal differences in the organization of their life history, habits or diet. Hibernation not only involves major changes in BMR but efficient rewarming entails associated adaptations such as enzymes that work at low temperatures and choosing the appropriate dietary lipids. Such differences adapt the species to their particular habitats and confer slightly different physiological capacities: there is no such organism as a basic 'standard' mammal.

A comprehensive study of an animal's activities in the wild can expose much sharper contrasts of ecologically important aspects of an animal's physiological capabilities than can laboratory studies by themselves. The most complete picture of what any particular species is actually capable of doing, and hence where and under what conditions it can thrive, is usually obtained by a combination of detailed observations of what it normally does in the wild, and laboratory experiments on the relevant physiological systems. However, there is no absolutely reliable way of distinguishing general from species-specific features, and it is important to remember that we might have a different concept of the 'normal' properties of a 'typical' mammal if species other than rats and mice had been adopted as standard laboratory animals.

Objectives for Book 1

When you have completed this Book, you should be able to:

1.1 Define and use, or recognize definitions and applications of the **bold** terms.

1.2 Outline the theoretical basis and practical methods for measuring the rate of use of energy at rest and during exercise.

1.3 Explain the effects of a cold environment and rapid growth or healing on basal metabolic rate in different species.

1.4 Describe how the kind of fuel utilized can be measured in living subjects and explain how such data combined with careful measurements of energy expenditure can reveal complex relationships between activity, diet and metabolism.

1.5 Outline major steps in the research necessary to identify and establish the role of a hormone or cytokine, using examples from modulators of the metabolism in white adipose tissue.

1.6 Describe how leptin was identified and outline its known physiological actions.

1.7 Explain the significance of the chemical composition of storage lipids for wild animals.

Questions for Book 1

(Answers to questions are at the end of the Book.)

Question 1.1 (Objective 1.2)

Which of the following statements are *true* of heat production and utilization in animals?

(a)   Heat is necessary to power metabolic processes.

(b)   Nearly all metabolic processes produce some heat.

(c)   Mammals and birds are homeothermic because they have specialized heat-generating tissues.

(d)   A fixed proportion of the chemical energy that an animal obtains from its food is directed towards heat production.

(e)   Physiological heat production stops if there is sufficient external heat to warm the body, thereby reducing the animal's need for food.

(f)   Most metabolic processes proceed faster at higher temperatures.

Question 1.2 (Objective 1.3)

(a)   Explain how choosing a cool place in the daytime promotes the deposition of storage lipid in bats.

(b)   Why does the body temperature of rats fall only slightly even when they are starving?

Question 1.3 (Objective 1.4)

Which of the following physiological processes can be monitored by measuring the respiratory exchange ratio?

(a)   anaerobic respiration

(b)   lipolysis and release of lipid from adipose tissue

(c)   utilization of fatty acids or glucose as fuel

(d)   heat production

(e)   deposition of lipid in adipose tissue

Question 1.4 (Objectives 1.5 and 1.6)

In a single paragraph, outline the kinds of observations and experiments that led to the discovery of the source of insulin and its action on adipose tissue. Why were such different techniques used in the discovery of leptin and cytokines such as tumour necrosis factor α?

Question 1.5 (Objective 1.6)

In a few sentences, outline the known properties of leptin production and reception that enable the hormone to control (a) appetite, (b) the mass of adipose tissue and (c) fertility.

Question 1.6 (Objective 1.7)

Why is the fatty acid composition of dietary lipids much more important to mammals that are fattening before entering hibernation than to birds (or mammals) that fatten before migration?

REFERENCES AND FURTHER READING

Döring, H., Schwarzer, K., Nuesslein-Hildesheim, B. and Schmidt, I. (1998) Leptin selectively increases energy expenditure of food-restricted lean mice. *International Journal of Obesity*, **22 (Suppl. 2)**: 83–88.

Emery, P. W., Bosagh Zadeh, A. R. and Wasylyk, A. (1999) The effect of malnutrition on the metabolic response to surgery. *British Journal of Nutrition*, **81**: 115–120.

Frank, C. L. and Storey, K. B. (1995) The optimal depot fat composition for hibernation by golden-mantled ground squirrels (*Spermophilus laterali*). *Journal of Comparative Physiology,* **164B**: 536–542.

Friedman, J. M. and Halaas, J. L. (1998) Leptin and the regulation of body weight in mammals. *Nature*, **395**: 763–770.

Geiser, F. and Kenagy, G. J. (1987) Polyunsaturated lipid diet lengthens torpor and reduces body temperature in a hibernator. *American Journal of Physiology*, **252**: R897–R901.

Geiser, F. and Kenagy, G. J. (1993) Dietary fats and torpor patterns in hibernating ground squirrels. *Canadian Journal of Zoology*, **71**: 1182–1186.

Green, A. and Newsholme, E. A. (1979) Sensitivity of glucose uptake and lipolysis of white adipocytes of the rat to insulin and effects of some metabolites. *Biochemical Journal*, **180**: 365–370.

MacQueen, H. A. and Pond, C. M. (1998) Immunofluorescent localization of tumour necrosis factor-α receptors on the popliteal lymph node and the surrounding adipose tissue following simulated immune challenge. *Journal of Anatomy*, **192**: 223–231.

Maehlum, S., Grandmontagne, M., Newsholme, E. A. and Sejersted, O. M. (1986) Magnitude and duration of postexercise oxygen consumption in healthy young subjects. *Metabolism*, **35**: 425–429.

Pond, C. M. (1998) *The Fats of Life.* Cambridge University Press, Cambridge.

Speakman, J. R. and Rowland, A. (1999) Preparing for inactivity: how insectivorous bats deposit a fat store for hibernation. *Proceedings of the Nutrition Society*, **58**: 123–131.

Stewart, J. M., English, T. E. and Storey, K. B. (1998) Comparisons of the effects of temperature on the liver fatty acid binding proteins from hibernator and nonhibernator mammals. *Biochemistry and Cell Biology*, **76**: 593–599.

ANSWERS TO QUESTIONS

Question 1.1

Only (b) and (f) are true.

(a) is false; heat is essential for raising the body temperature and thereby enabling metabolic processes to proceed smoothly and rapidly, but the heat itself does not contribute to the energy that fuels the reactions.

(c) is false; some mammals have a specialized thermogenic tissue (BAT), especially during hibernation and as neonates, but even in these animals more than 90% of the total body heat is generated by other organs.

(d) is false because the efficiency (and hence heat production) of many metabolic processes (e.g. muscle contraction) varies with the conditions under which they take place.

(e) must be false if (b) is true. Metabolism cannot be stopped totally except in death so heat production as a by-product of metabolic processes cannot be eliminated completely.

Question 1.2

(a)   Being in a cool environment enables the bat to cool down rapidly as soon as it has stopped feeding and become torpid. Metabolism continues, albeit more slowly at the lower temperature, but energy expenditure is much lower. The energy so saved enables more lipid to be deposited in adipose tissue.

(b)   Rats are unable to become torpid because they lack the biochemical adaptations to enable metabolism to continue at low temperature. They therefore continue to devote energy to keeping warm even when starving.

Question 1.3

(c) is the right answer. The others are wrong because (a) RER includes oxygen uptake which is not used in anaerobic respiration, (b) and (e) mobilization and deposition of lipids do not entail gas exchange, (d) RER does not measure temperature.

Question 1.4

Surgical removal of a single small gland, the pancreas, produced symptoms similar to the human disease, diabetes. Although ill, animals so treated could be kept alive for weeks. Insulin was isolated and purified from the pancreases of domestic livestock. Its regular injection into diabetic dogs cured the immediate symptoms of the disease.

These kinds of experiments cannot reveal the source or roles of leptin and cytokines because these substances are produced by many different tissues, occur at very low concentrations in the blood, and have multiple actions. Leptin was identified only after a mouse carrying a mutant gene for the hormone was studied, and modern techniques of genetic engineering could be used to produce it from bacteria in large quantities.

Question 1.5

(a)   Leptin circulates in the blood and acts, probably via other messenger molecules, on the hypothalamus which controls feeding behaviour. Rodents that cannot produce leptin or lack leptin receptors have huge appetites and, over several weeks, grow obese.

(b)   Leptin is secreted from adipocytes of at least the larger depots of adipose tissue. It is believed to signal lipid reserves because depleted adipocytes and those of lean animals secrete larger quantities of leptin than replete adipocytes in recently fed animals.

(c)   Homozygous *ob/ob* mice are infertile, though this deficiency can be corrected by regularly injecting leptin. Artificially supplementing blood leptin levels in juvenile mice hastens sexual maturation. Leptin and leptin receptors are produced in male and female reproductive tissues.

Question 1.6

Mammals and birds can oxidize any of a wide range of non-toxic fatty acids as fuel for activities such as migration that take place at their normal (warm) body temperature. Although the carrier proteins and enzymes are adapted to function in the cold in animals that normally hibernate, only certain fatty acids, particularly long-chain polyunsaturates, can be oxidized efficiently under such conditions.

ACKNOWLEDGEMENTS

Grateful acknowledgement is made to the following sources for permission to reproduce material in this book:

Cover: Courtesy of Dr David Robinson; *Figures.1.1, 1.2*: Speakman, J. R. and Rowland, A. (1999) Preparing for inactivity: how insectivorous bats deposit a fat store for hibernation, *Proceedings of the Nutrition Society*, **58**, CABI Publishing; *Figure 1.3*: Maehlum, S., Grandmontagne, M., Newsholme, E. A. and Sejersted, O. M. (1986) Magnitude and duration of excess postexercise oxygen consumption in healthy young subjects, *Metabolism*, **35**(5), Grune and Stratton Inc.; *Figure 1.5*: Green, A. and Newsholme, E. A. (1978) Sensitivity of glucose uptake and lipolysis of white adipocytes of the rat insulin and effects of some metabolites, *Biochemical Journal*, **180**, The Biochemical Society and Portland Press; *Figure 1.7*: Doring, H. *et al.* (1998) Leptin selectively increases energy expenditure of food restricted lean mice, *International Journal of Obesity*, **22**, Stockton Press/Macmillan Press Limited; *Figure 1.8*: Geiser, F. and Kenagy, G. J. (1993) Dietary fats and torpor patterns in hibernating ground squirrels, *Canadian Journal of Zoology*, **71**, National Research Council of Canada.